PARTS OF A CELL

For Curious Little Minds

Look at all the snowflakes.
Let's make a snowman.

Snowflakes are the basic
building blocks we need to
make our snowman.

Wait! What am I made of?
Cells. Just like a snowman is made of snowflakes, you are made of cells.

What are cells?

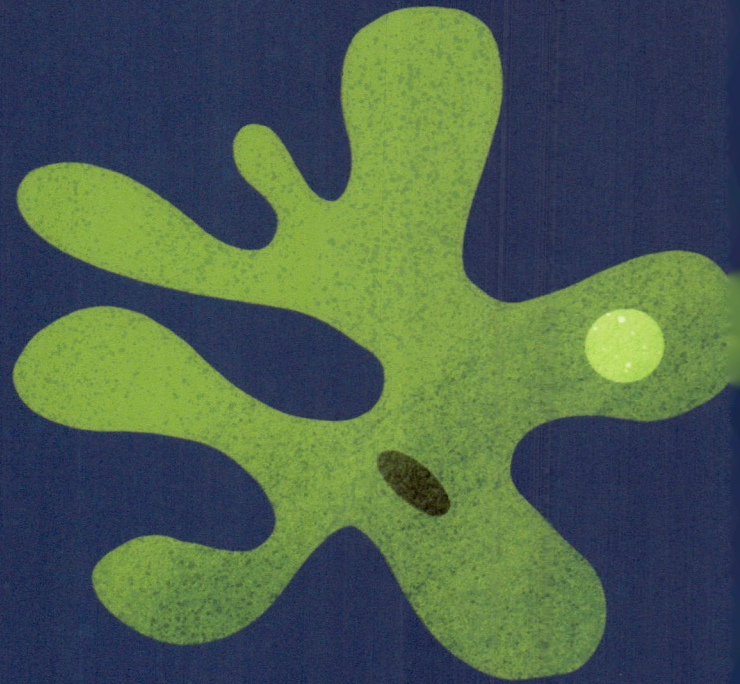

Single Cells

Cells are the basic building blocks of all living things. Some are single-celled like the Amoeba.

Some are multi-celled like humans. Unlike snowflakes, cells are so tiny that we need a special tool to see them. It is called a microscope.

Multi-Cells

Brain Cells
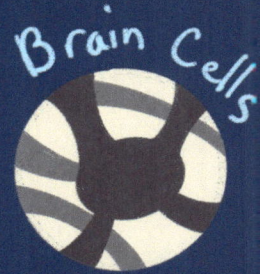

Nerve Cells
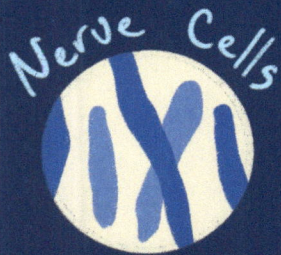

Red Blood Cells

Fat Cells
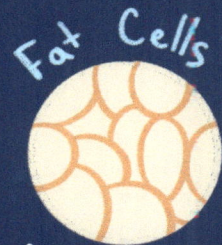

What is a mi.cro.scope?

A microscope helps us zoom into very small things. It is like a magnifying glass but more powerful.

Using a microscope we can peek inside a cell and see that it is made up of many smaller parts called organelles.

LLES
Organelles are like little helpers that are found inside a cell and have specific jobs that help the cell live. Let us learn about these organelles by drawing them.
Ribosomes
Microtubules
Yeah! I love drawing!

Cell Membrane

Parts of a Cell

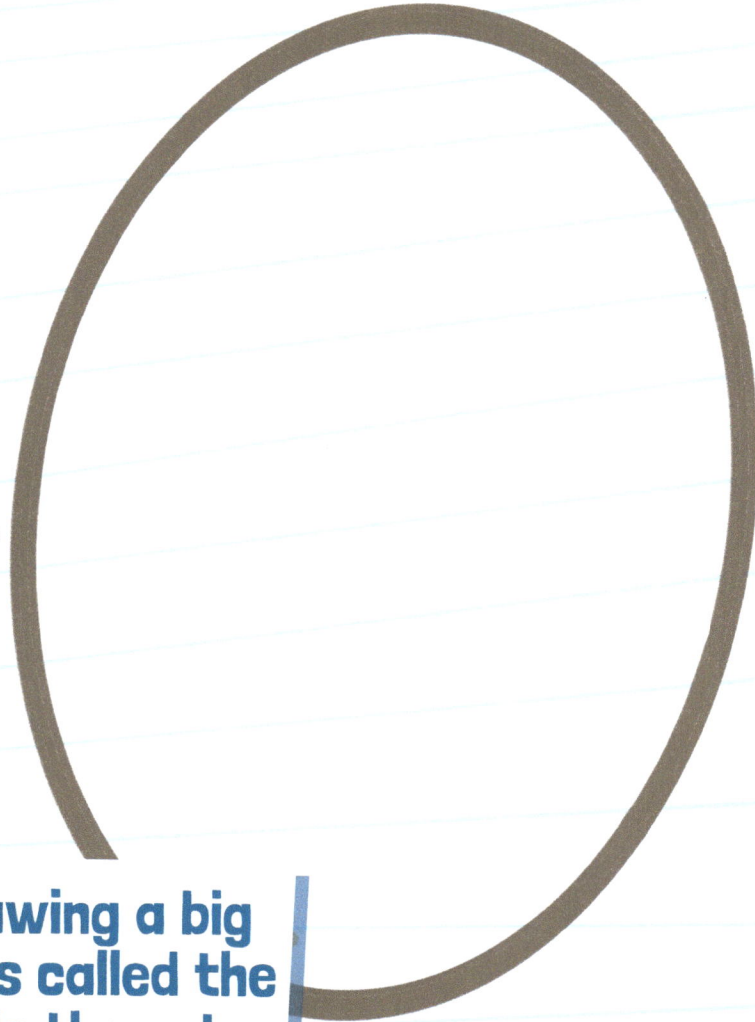

Let's start by drawing a big brown circle. This is called the cell membrane. It is the outer structure of the cell.

Nucleus

Parts of a Cell

Now draw a red circle in the middle. This is called the nucleus. It is the brain of the cell. It uses DNA to control cell activity.

Ribosomes

Parts of a Cell

Now draw some small yellow dots. These are called ribosomes. They use the DNA to make proteins.

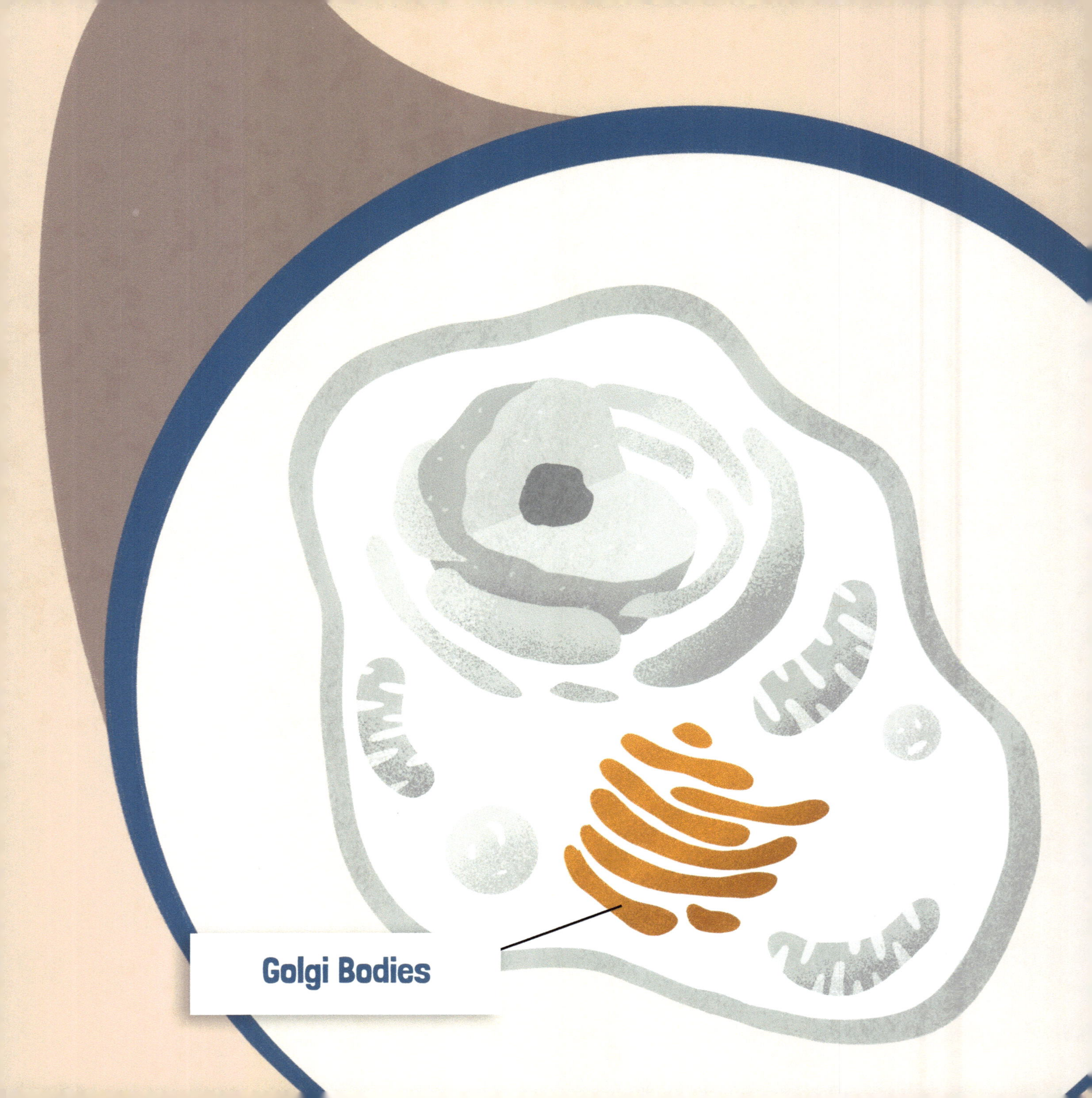
Golgi Bodies

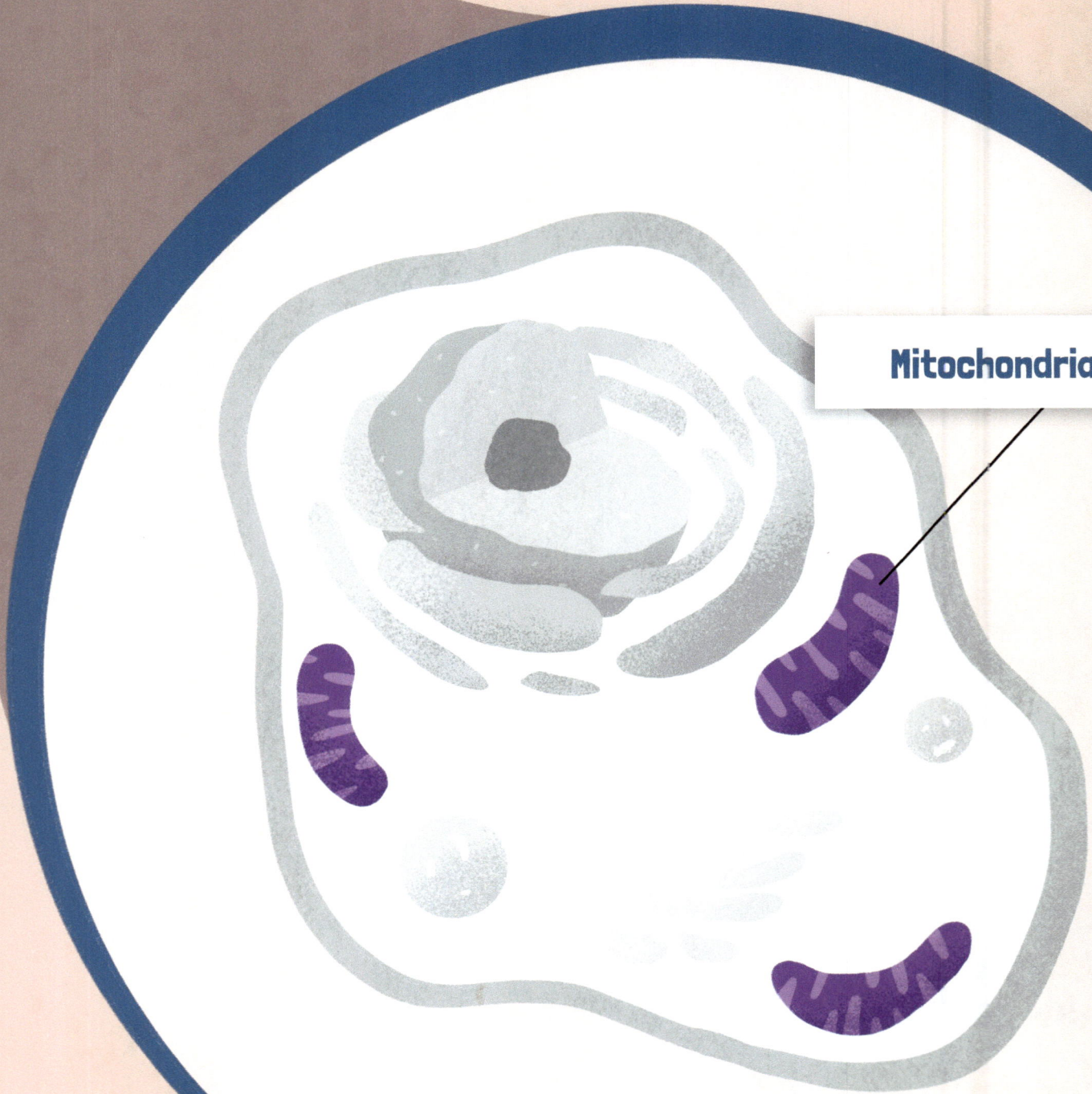

Mitochondria

Parts of a Cell

Now draw some purple ovals. These are called Mitochondria. They produce energy for the cell.

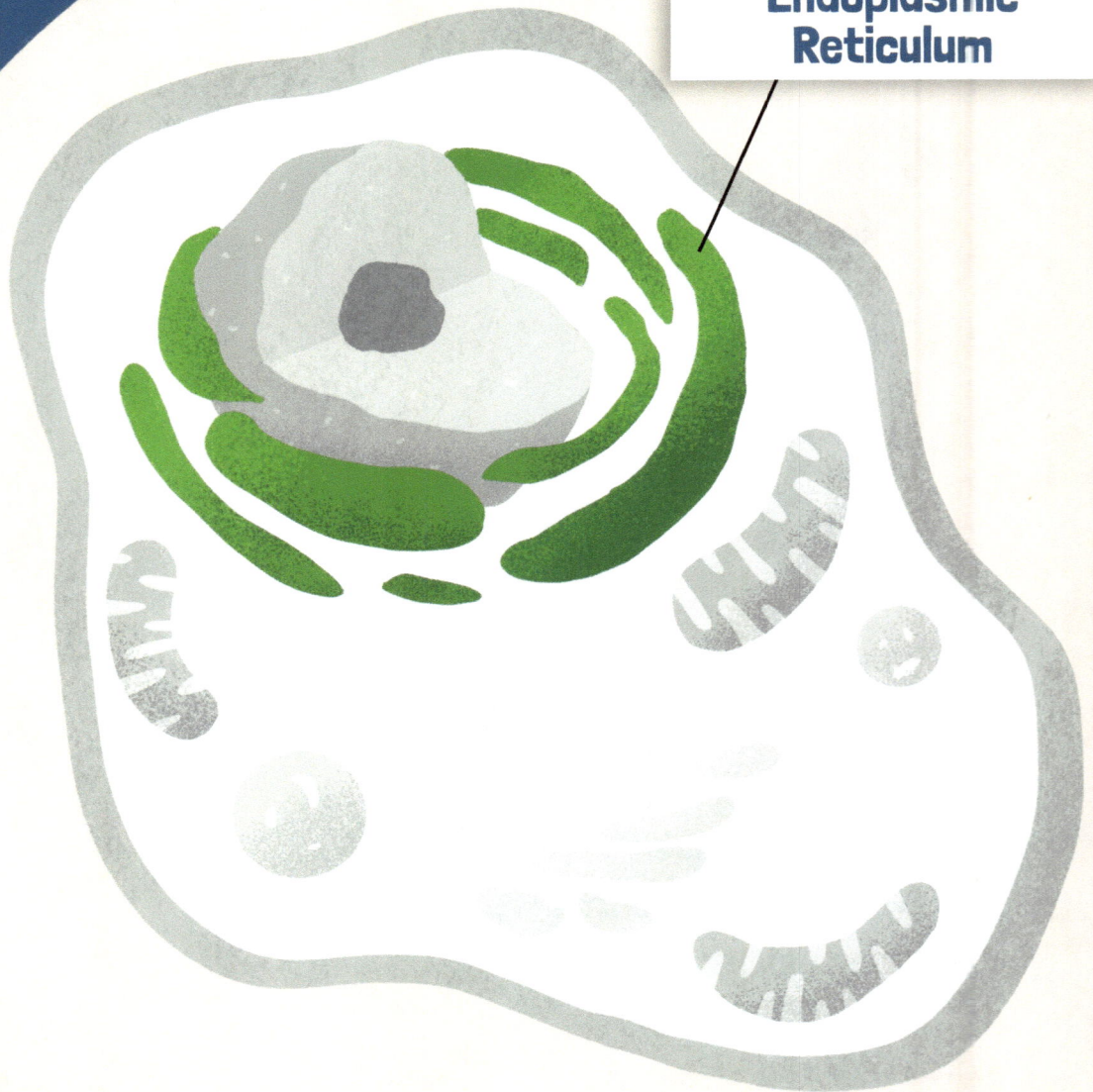

Endoplasmic Reticulum

Parts of a Cell

Now draw some green squiggles around the nucleus. This is called the Endoplasmic Reticulum. This is the place for making proteins.

Lysosomes

Parts of a Cell

Now draw some small pink circles. These are called lysosomes. They recycle materials and remove wastes.

Cytoplasm

Parts of a Cell

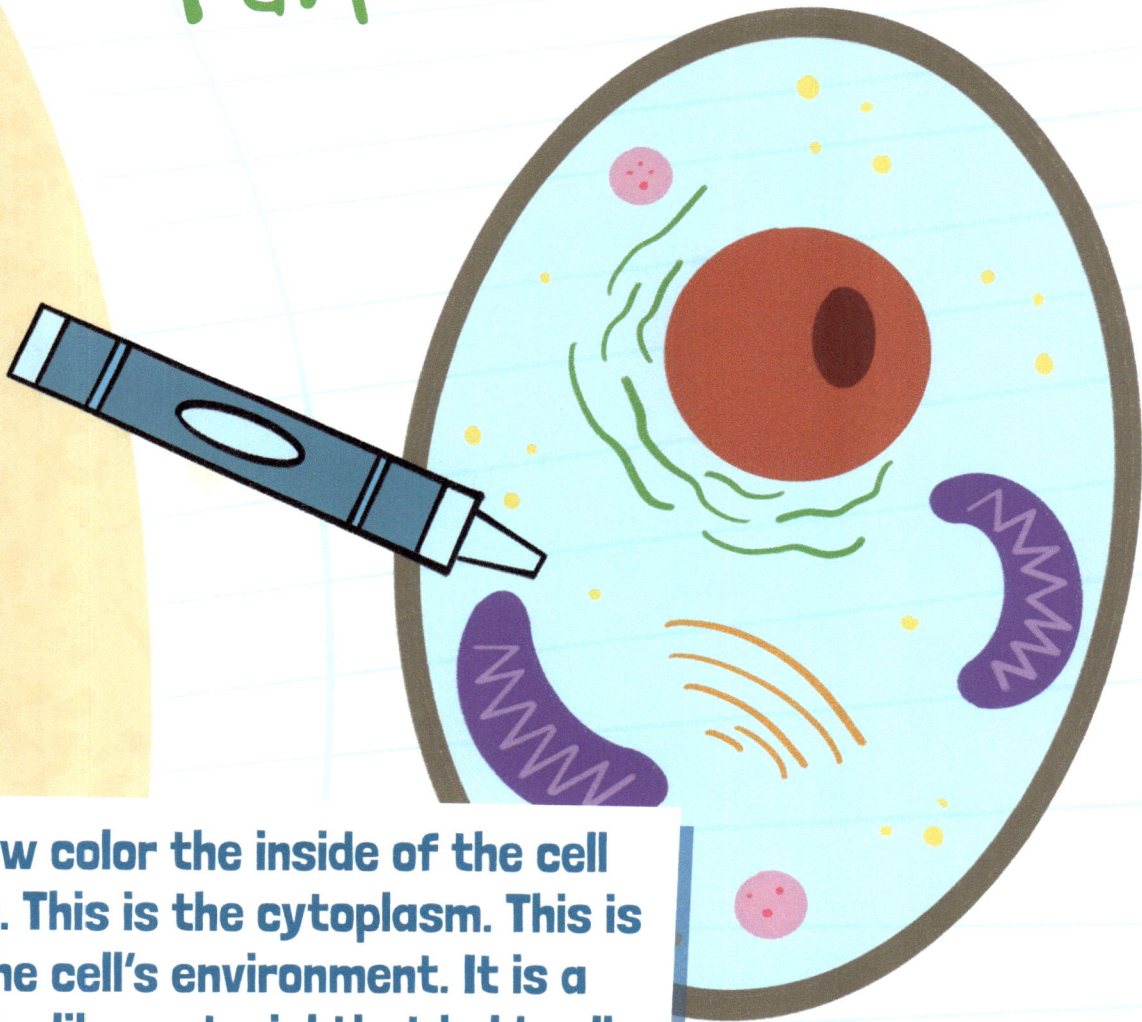

Now color the inside of the cell blue. This is the cytoplasm. This is the cell's environment. It is a jelly-like material that holds all the organelles in place.

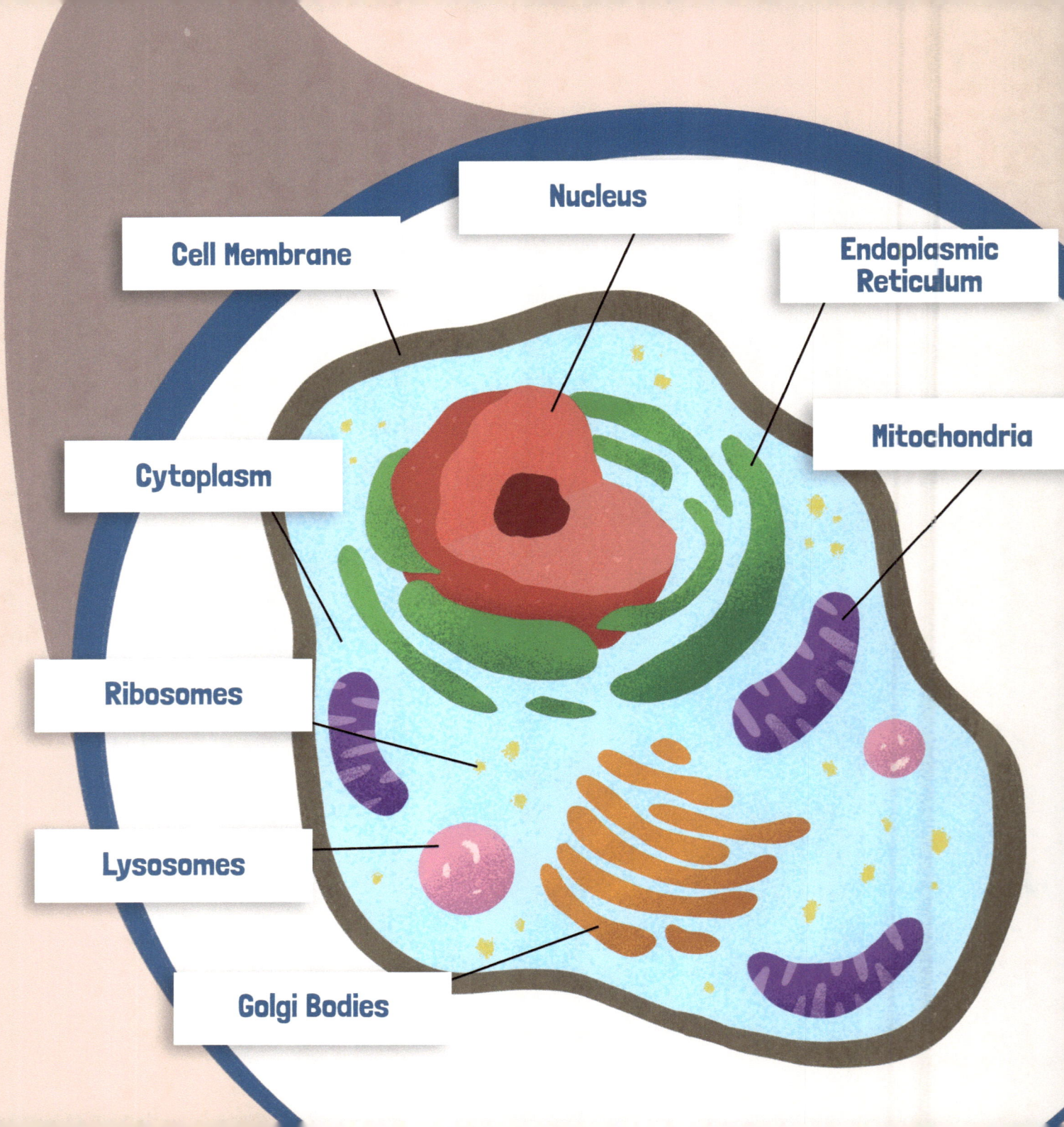

Cell Membrane

Nucleus

Endoplasmic Reticulum

Mitochondria

Cytoplasm

Ribosomes

Lysosomes

Golgi Bodies

Parts of a Cell

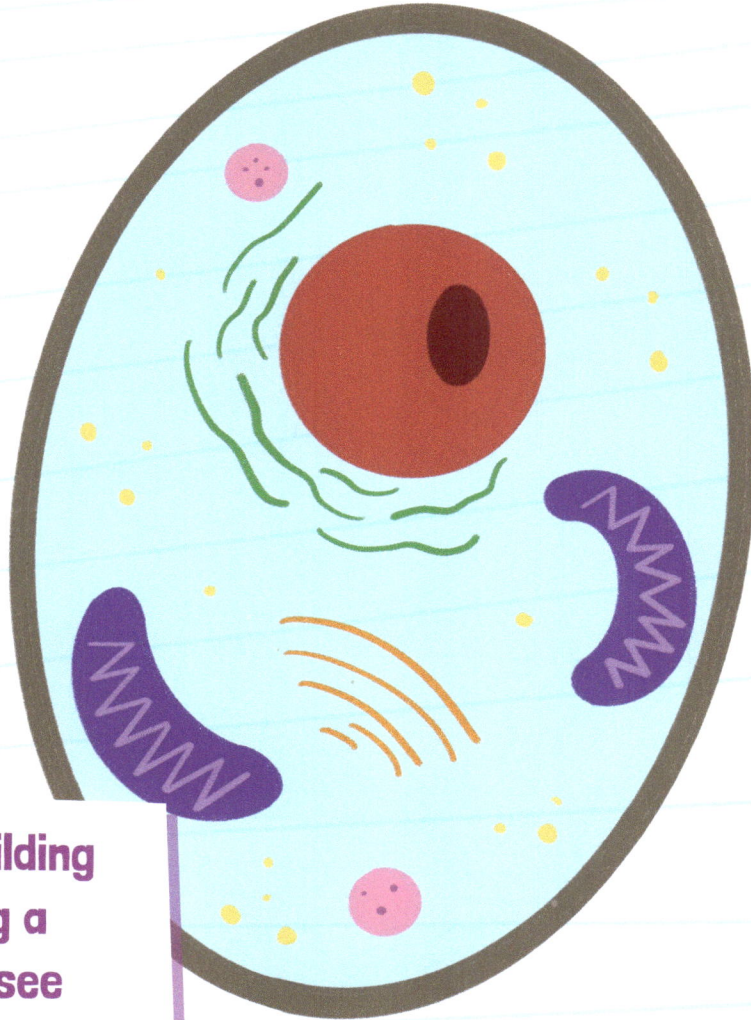

Cells! They are the building blocks of life. Using a microscope we can see the different parts of a cell called organelles.

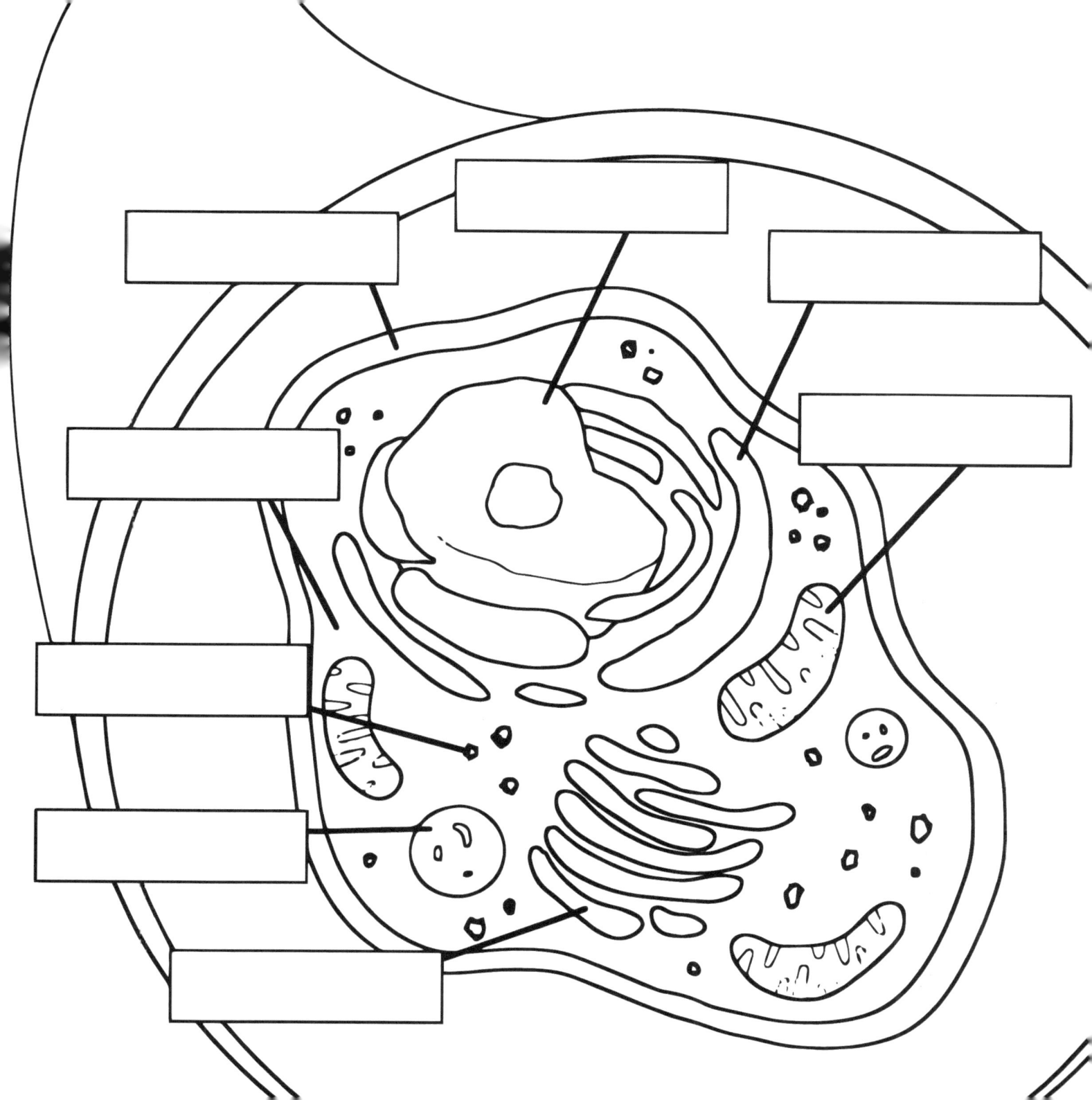

PARTS OF A CELL

For Curious Little Minds

www.ingramcontent.com/pod-product-compliance
Lightning Source LLC
Chambersburg PA
CBHW040022050426
42452CB00002B/92